Reaseheath

Picture of a college

Written and compiled by Judith Dooley

Dedication

Peter R. Dale B.A., A.L.A. (1950-2003)

Peter Dale was the only son of Raymond and Dorothy Dale. He was educated at Newcastle High School and Aberystwyth University where he gained his degree in Geography and his post-graduate qualification in Librarianship.

From 1974 until 2002 when he was stricken with cancer he was Librarian at Reaseheath College. He established and built up the library into one of the most comprehensive collections of specialist knowledge in the country.

Peter was a keen golfer and a lifetime supporter of Stoke City Football Club. His prodigious general knowledge fuelled his passion for quizzes, both as a compiler and a competitor.

Married to Jane, he was the father of two sons, Owen and Alex.

Staff and students of Reaseheath College remember Peter with affection and gratitude.

Introduction

Cheshire School of Agriculture opened in 1921 when George V was King and Lloyd George was Prime Minister. Women had to be over thirty in order to vote and it would still be another seven years before the invention of penicillin. The celebrated young film star of the day was Charlie Chaplin. The countryside was still largely farmed in traditional ways: the rural scene one of horses pulling the ploughs and farm labourers toiling from dawn until dusk largely without the aid of machines.

It was a very different world from the one we know today, and the changes and advances made by science and technology in the intervening years are greater than at any other time in history.

Cheshire School of Agriculture grew and developed with the changing times. It became Cheshire College of Agriculture and then Reaseheath College and throughout the years continued to rise to the challenge of providing up-to-the-minute specialist education to generations of students.

This pictoral history shows Reaseheath as it was and as it is, and illustrates why the college has always inspired pride and affection in both the educational world and the local community.

Growing
the curriculum

When the Cheshire School of Agriculture at Reaseheath opened to male students in 1921, the affiliated Worleston Dairy Institute had already been training girls in the ancient craft of cheesemaking since 1892. The number of students at the college in the twenties was only around sixty, including both male students and female dairy students. Dairying was transferred to the main campus in 1926, when the women's hostel was built. The Principal during these early years was W.B.Mercer.

From 1921 to the outbreak of the Second World War in 1939 the college provided one-year certificate courses in agriculture, poultry husbandry, dairying and horticulture. These courses were suspended during the war years and when they were reintroduced after the war it was initially in a shortened form.

By 1947, under the new Principal, J.K.Lambert, the full range of coursework was once more available and began to expand from the one year certificate to include short courses such as farmhouse cheesemaking and the two year course in Technical Education for the Dairy Industry. Outreach, day release

and evening classes in Agriculture and Horticulture were also introduced.

George England became Principal in 1966, a time of rapid technological advances in the field of land based industries, to which he responded by implementing expansion of the curriculum and by developing the practical resources needed to keep the college at the forefront of this type of specialist education.

By 1987, when George England retired and Vic Croxson became Principal, the college had grown in both size and diversity. There were now 265 full time students, 1,425 part time students, a staff of 130 and a budget of £2.4 million. The Cheshire College of Agriculture, as it was now known, remained part of Cheshire County Council's education provision. There were four main study areas: agriculture, food and dairy science, engineering and horticulture. Forty percent of the full time students at this time studied farming courses.

In 1993 Reaseheath College was incorporated, along with all other colleges of further education, by Act of Parliament. All assets were transferred from the Local Education Authority and were vested in a Board

that assumed full responsibility for the affairs of the college. Funding was allocated by a new government quango – the Further Education Funding Council – according to an increasingly complex formula which made life financially challenging for many similar specialist colleges.

The Reaseheath College Board, led by John Platt D.L., O.B.E., A.R.Ag.S., pursued a strategy of maintaining independence through the diversification of college activities and the development of on-site partnerships with organizations that complement the college's mission.

The modern college is a dynamic and thriving business. Student numbers have reached 900 full time and 5,500 part time in 17 programme areas. Staff numbers exceed 300 and the annual budget is now £9million. The number of students studying farming, however, has fallen to eight percent, a reflection of the way modern agriculture in Britain is now structured.

A group photograph taken circa 1930. The dog belongs to the Principal, W.B.Mercer, and is called Tweed. Apparently Mr. Mercer always named his dogs after Scottish rivers.

An early photograph of Worleston Dairy Institute, which later became part of the college. This group of student dairymaids was pictured in 1914 with their Principal, Miss J. Forster N.D.D.

Left: Dairy students in the 1930s, when the Dairy Teaching Unit had become part of the Cheshire School of Agriculture. These girls are seen turning the cheeses as part of the cheesemaking process.

Right: Lecturer Douglas Taylor with a class of horticulture students in the 1950s.

Below: This lovely old photograph of the spraying of fruit trees was taken in the early years of the college before Health and Safety became the issue that it is today.

Continuing care for the trees - decades later lecturer Geoff Scaife hanging from the branches in the late 1980s.

Two women pictured in the dairy in the 1930s, immaculate in their white overalls.

Blowing in the wind – enjoying a lesson out of doors, this group of students in the 1970s is gathering plant samples.

An integral part of
farming is boundary
security. This group of
agricultural students is
pictured in the late 1970s
constructing a fence.

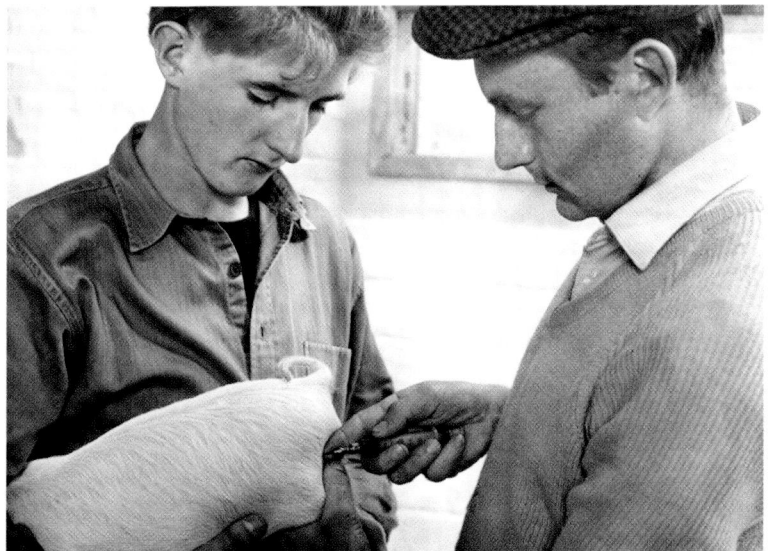

The curly tail indicates a
pig on the receiving end of
an injection by two
students in the 1970s.

Top left: A group photograph taken outside Reaseheath Hall in 1936, female members of staff resplendent in their hats and the dairy students conspicuous in their white aprons and caps.

Below left: Although three decades separate this group photograph, taken in 1967, from the previous one, it can be seen from the smart attire of the students that certain standards in dress were still adhered to.

Principal George England enjoys teaching time with a group of attentive students out in the fields.

The start of the college library, pictured in the 1970s. Peter Dale is second right in the background of the picture.

The modern college library, equipped with study areas both downstairs and up. This library was the work of Peter Dale, who built it up to meet the needs of the expanding college.

Modern times see information being accessed from more than books. These students are working in one of the college computer suites finding what they need on the Internet.

Potting off – two well dressed students from the early 1950s remind us of a time when smartness was expected in educational institutions, unlike the more casual approach to dress of the modern student. The college had a dress code for meals up until 1973.

Right: These flowerbeds in front of Reaseheath Hall are one of the first impressions a visitor to the college receives. They are created and maintained by the Horticulture students such as this group pictured in the 1990s.

This pretty garden has now disappeared to make way for the Lord Woolley Centre. Andrew Lamberton, son of Principal J.K.Lamberton, is pictured here with his dog.

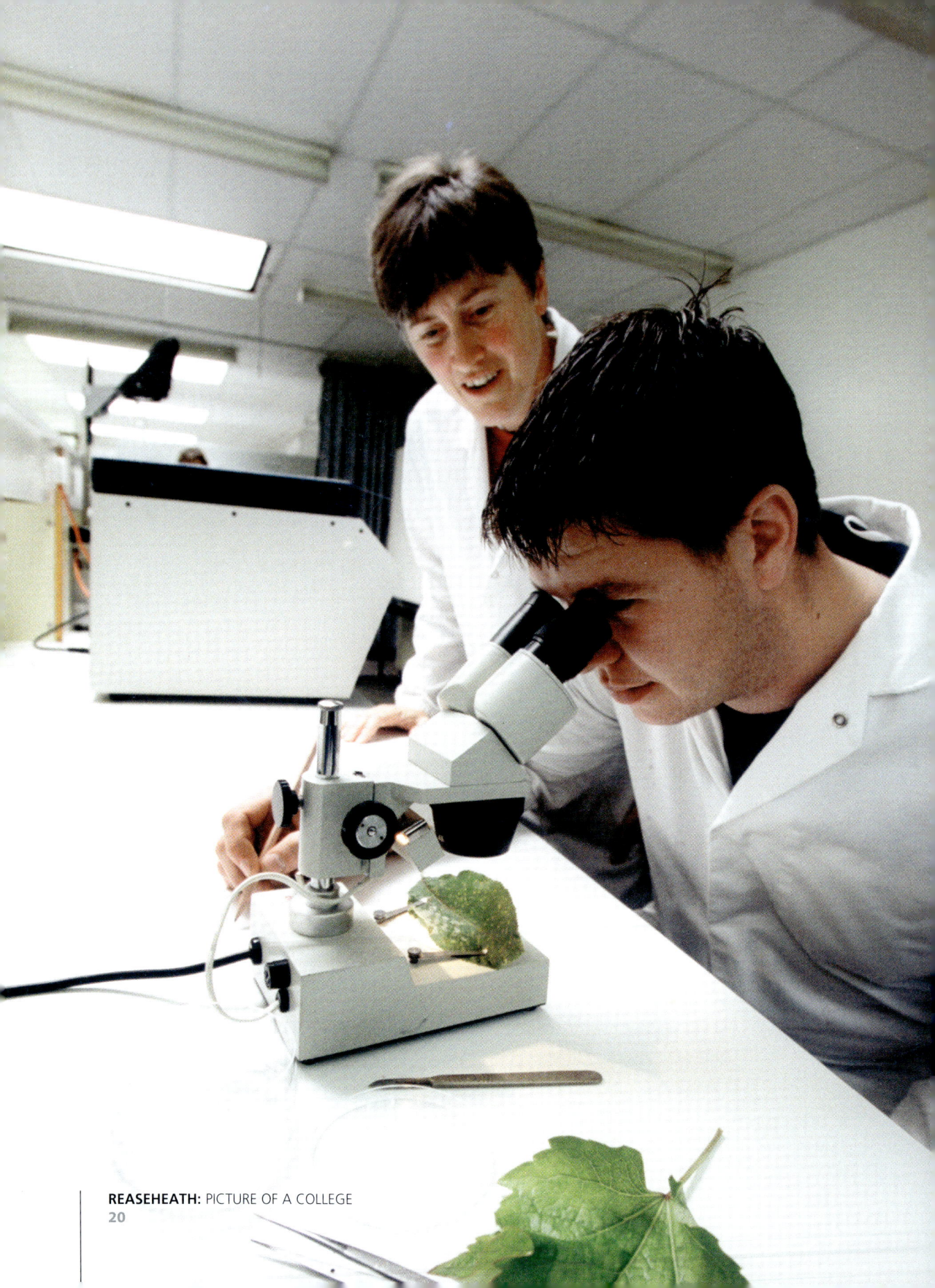

A poultry husbandry class in the 1930s (top right) is a far cry from plant science in the 21st Century (left). But though many years and different topics separate these photographs it can be seen that the basic principles of study have remained the same.

Making a point – a lecturer giving instructions to one of his students on a Construction course.

Sheep shearing must be one of the most ancient tasks known to man, practised since animals were first kept for domestic use. Here a practitioner of the 21st century performs this traditional task.

Over the years Reaseheath has kept up a steady building expansion and improvement programme to house its activities. This picture shows the extensive restoration of the farm office in Spring 1988.

One of the most popular sectors of the college is Animal Care. This student has her Labrador well under control.

Right: Early students at the college could never have imagined that this would be a photograph taken on a college course. It shows Adventure Sports students enjoying the river.

On top of the world.
Adventure sports lecturer,
Jon Mercer.

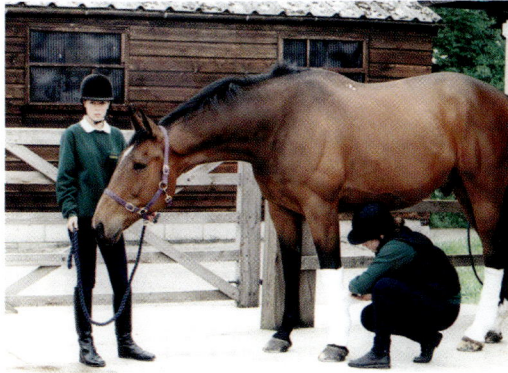

Equine students check the bandages on this horse's legs.

In the late nineties the college widened its intake by providing courses from a Pre-Entry level. This student is on a Supported Learning Programme where courses are designed to improve basic skills and meet special needs in a challenging practical environment.

The milking shed in the nineteen forties. This young man wears a collar and tie when filling in the records on his clipboard. The milking shed with its row of buckets makes an interesting contrast to the following picture.

Above right: Milking moves with the times – this is the milking parlour at Reaseheath pictured in the 90s.

Below right: The milking parlour is also a teaching resource as students can watch operations from this viewing gallery.

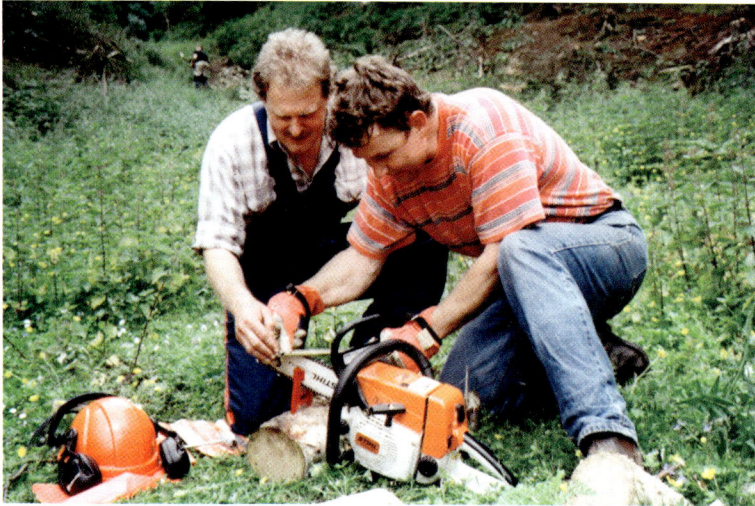

Health and Safety is closely
monitored at the college.
The advice given by
lecturer Steve Roach to this
student is designed to
prevent serious injury.

Made to last – painstaking
and time-consuming work
by Countryside students
goes into the construction
of these steps.

Equally painstaking is the creative work put into these delightful floral arrangements being made by this group of floristry students.

A tranquil scene from the past – Joe Adlard is pictured with Dolly the horse, cleaning Reaseheath lake in the nineteen twenties.

Sophisticated and powerful farm machinery is an integral part of modern agriculture and students are trained in its use and maintenance.

Sparks fly as this student in agricultural engineering carries out repairs.

In the early days of the college ferrets would have been working animals used to catch and kill rabbits. Nowadays they are kept as pets and students learn how to handle and care for them as part of their Animal Care course.

Pictured below is a student with one of the more exotic residents of the Animal Care Unit.

Right: Into the new millennium: Equine Studies is an increasingly popular area of study and the college boasts a purpose built Equine Unit.

Scattering expertise

Reaseheath College has always been justly proud of the quality of its specialist education in agriculture and the land-based industries, but its influence has been more far reaching than simply the education of its students.

One of the earliest examples of the college's contribution to the community was in the Second World War when recruits for the Women's Land Army came to Reaseheath for training in the agricultural skills needed to keep up production of essential food supplies for Britain.

Set in the heart of rural Cheshire the college enjoys a role within the local agricultural community and is closely concerned with organizations such as the Cheshire Federation of Young Farmers, the Cheshire Agricultural Society, the Cheshire Grassland Society and the Cheshire Farms Competitions.

Small children from local schools love the experience of the Greenway Trail and older children enjoy visits to the various departments of the college at a time when they are considering further education choices.

The name of Reaseheath College is also known in many countries abroad, including Egypt, Romania, Zimbabwe, USA, Uzbekistan and SaudiArabia, where lecturers from the college have passed on their knowledge and expertise. Today staff from Reaseheath still teach dairying skills all over Britain.

At the outbreak of the Second World War the college closed down for the duration and became a training centre for the Women's Land Army and the base for the War Agricultural Committee, pictured here on the lawn at Reaseheath Hall.

Girls in the Women's Land Army in the Horticulture Department of Reaseheath College, then Cheshire School of Agriculture, where they were trained in the techniques of market gardening.

Left: Girls were trained in the feeding of cows and calves and in milking both by hand and machine. Many farms did all the milking by hand but at Reaseheath one shippon was entirely given to machine milking.

By 1940 thousands of men had left the farms to fight with the British Armed Services and Britain needed to start producing all its own food. Growing wheat and potatoes was especially important. The amount of land used for crop growing in Britain increased by 50% during the war.

This bull could have been very daunting to the girls who trained at Reaseheath, who were all "townies" and therefore raw recruits. Country girls could usually be placed on a farm immediately. The girls at Reaseheath received four weeks training, working twelve-hour days of gruelling hard work, including harvesting and haymaking.

Above: A verse from the Women's Land Army song: "Back to the land, we must all lend a hand. To the farms and the fields we must go. There's a job to be done, Though we can't fire a gun We can still do our bit with the hoe. Back to the land with its clay and its sand, Its granite and gravel and grit. You grow barley and wheat And potatoes to eat To make sure that the nation keeps fit. We will tell you once more You can help win the war If you come with us back to the land."

These empty baskets will be much harder to carry when filled with fruit and vegetables. Despite the long hours and hard work the girls had a sense of camaraderie and often remained friends for life.

These sheaves of wheat would have been cut by a horse-drawn "self-binder", the wheat dropping onto a moving belt with a sheaf tying mechanism on the side. The sheaves were then propped up in stooks to dry before being loaded on to a cart and pulled by horses to the haystack. This Landgirl is passing sheaves to the top of the stack. When finished the stack would be thatched against the weather.

Left: Clearing out the hedges and burning the rubbish. The work done by the Women's Land Army showed that women were capable of performing tasks that had traditionally been men's work, and this contributed towards improving women's position in society.

Jack Hamlyn demonstrates the use of a "scuffle" to attentive Landgirls. This implement, harnessed to a horse and guided by the handles at the back, was used to remove weeds from between the rows of green crops such as swedes, mangels, turnips and kale.

Bert Davies " the apple man" pictured during one of his annual apple events. Bert is a top fruit expert who approached Reaseheath College with the suggestion of setting up a small fruit garden where amateur gardeners could come for advice. The garden has been running successfully since 1991.

Reaseheath lecturer Simon Young pictured in the 1980s in a school hall with an audience of entranced children – his talk about farming and the countryside is illustrated with live chickens and farm animals.

In January 1991 work started on the Greenway Trail to provide a three mile long walk through the grounds of the college, giving visitors, especially children, safe and easy access to the delights of the countryside.

The Greenway Trail evolved into three walks of different lengths exploring various areas of Reaseheath. Over the years thousands of visitors have trodden its paths and many small children have come with teachers and parents to enjoy sights such as this one in the farmyard.

Milk Marketing Board Home Economics Advisors learn how to make butter in Reaseheath Dairy with Ron Lawton, Head of the Dairy Department in the mid 1970s.

A community project in Uganda organized by Dig Woodvine, Managing Director of XCL Ltd, an expedition specialist company based at Reaseheath College. Retired Reaseheath arboriculture lecturer, Geoff Scaife, advised on producing timber as a cash crop. The team also carried out practical building projects in schools and on national park trails and trained young people in sports skills.

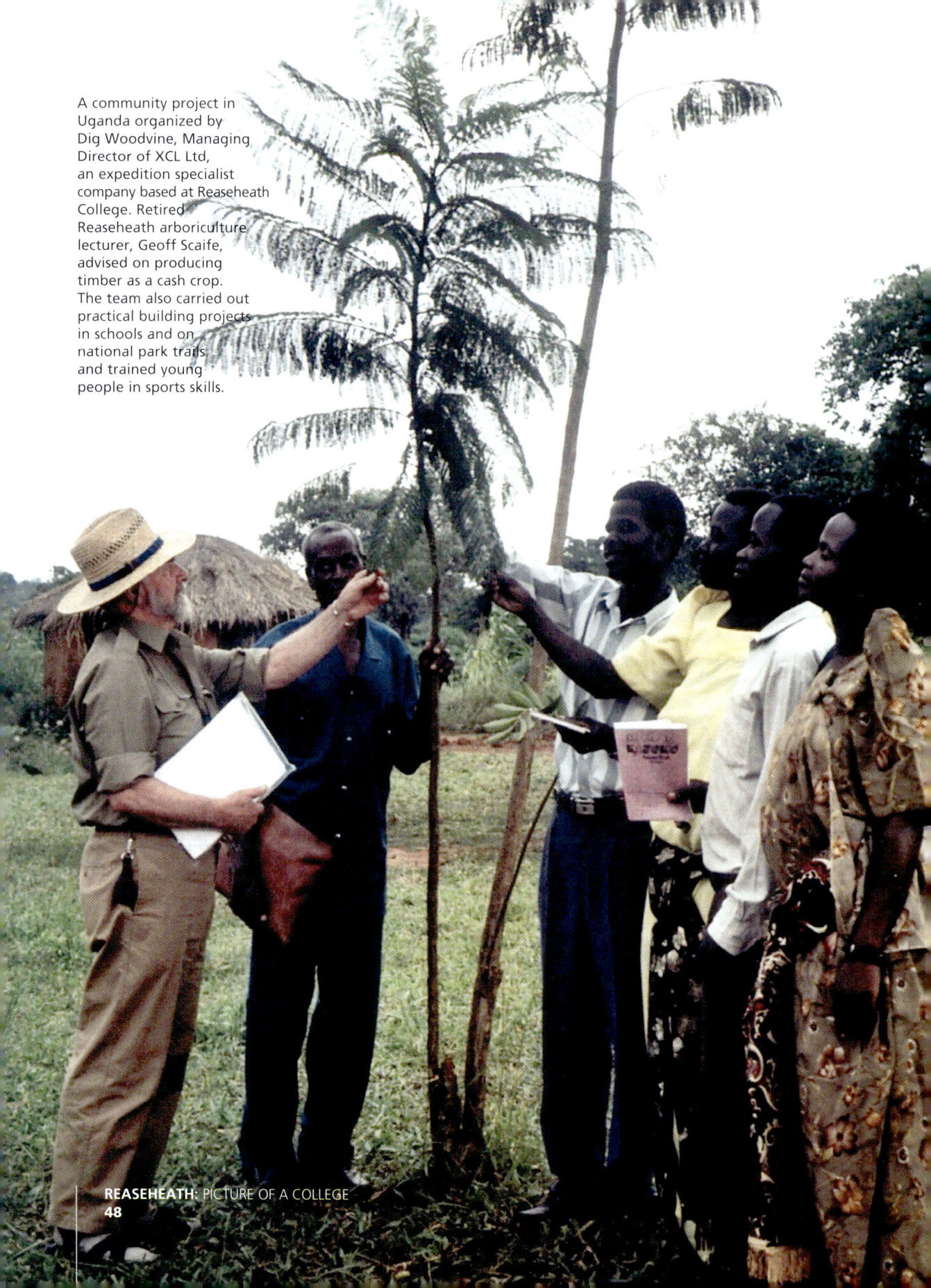

Taking a break in Saudi Arabia on the edge of the desert, Reaseheath lecturers and staff from Riyadh Agricultural University in the mid 1990s. They were working together to carry out an assessment for a feasibility study on milk products processing and marketing.

Reaseheath lecturer Dennis Prew, left, in Romania with a staff member from the Romanian University in the late 1990s.

Reaseheath lecturers visit Damanhur in Egypt. This discussion concerns equipment required for process halls for manufacturing milk products.

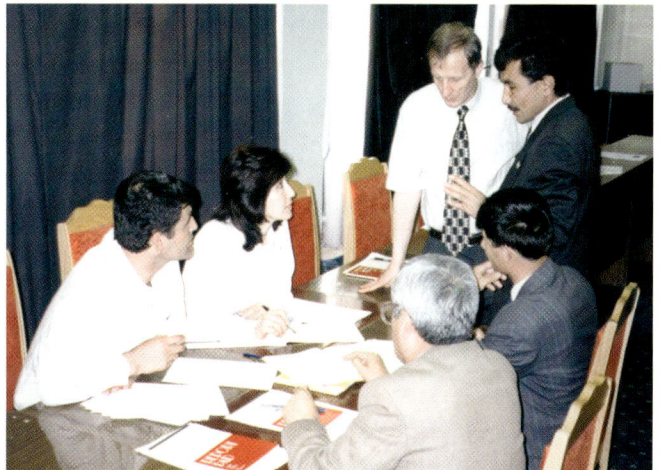

Dennis Prew in Uzbekistan, working with university staff on curriculum development for a course at Tashkent University.

In the early 1990s Manchester United Football Club contacted Reaseheath to ask if staff from the college could provide training for their groundstaff in the driving, handling and maintenance of ground machinery to meet the requirements of the Health and Safety Executive. Steve Roach and Dick Coultas (pictured here) visited Old Trafford and met with the Health and Safety Executive and Sir Alex Ferguson. They then planned and delivered a training programme to the 30 strong team who looked after the Old Trafford ground and the training sites of Manchester United.

Branching out

Reaseheath College now sits squarely in the world of modern business and as such uses its many attractions to generate money for reinvestment.

From the small agriculture-oriented establishment of the 1920s has grown the multi-faceted institution that we see today.

The facilities of the college are enjoyed by diverse groups from businesses holding meetings and conferences, holiday operators and regular visitors to the college shop and Garden Centre.

The Springtime lambing events are eagerly awaited every year and attract visitors from a wide radius. This is a time particularly enjoyed by children.

Another keen favourite with families is the annual Maize Maze, with a different theme each year, providing excitement and fun.

The college has its own thriving Golf Club with a course cared for by the Golf Course Management students. Another sporting connection is the Crewe Alex training ground on the fields bordering the Chester road.

As well as being an attractive venue for catered parties the college is now a firm favourite for weddings. The college has a license as a venue where the marriage ceremony can be performed and the Main Hall lends itself beautifully to such events. It woud be difficult to find anywhere providing such a beautiful natural backdrop for photo opportunities than the college gardens and lake.

The garden centre stocks a wide range of attractive plants, pots and garden products. There is always a good choice of plants for both indoors and the garden, and, of course, expert advice is on hand.

Reaseheath shop and garden centre is a commercial concern run by the college. Students use the shop for a variety of products ranging from fresh sandwiches and snacks to stationery.

The shop and garden centre also provide a useful teaching resource, stocked and maintained by Horticulture students.

Reaseheath Hall, as well as being a specialist education institution, is also available for conferences, meetings and exhibitions. The Conference and Banqueting staff also organize parties and weddings. A variety of rooms are used for outside purposes, including the main hall, which is pictured here. In the early days of the college huge fires used to burn in this fireplace and the hall was used for weekly dances by the students in what was an informal and almost family setting.

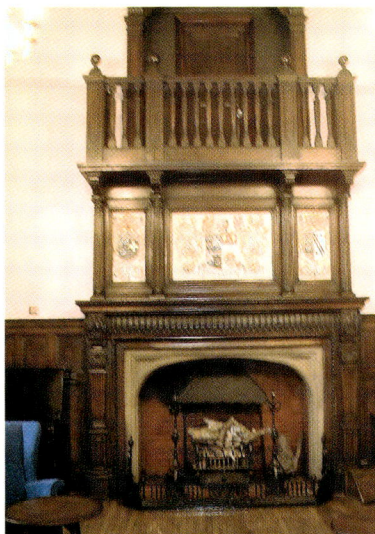

Reaseheath Hall and gardens are proving a popular venue for weddings and receptions. The marriage ceremony is performed in the main hall which, with its stained glass windows and sweeping staicase, lends itself well to such occasions.

Below: Sunshine, roses and champagne – the beautiful landscaped gardens of Reaseheath are perfect for those special wedding photographs.

Left: Some of the Reaseheath fields along the Chester road are now the training ground for Crewe Alexandra Football Club and supporters can often be seen watching their team practising.

Pictured at the Reaseheath training ground is Crewe Alexandra's team manager, Dario Gradi, in November 2001, the day before his 1000th game in charge of the team.

Reaseheath Golf Club was founded in 1987 with a nine-hole course which had been developed over several years as a practical teaching facility for students of green-keeping and golf course management. The club now has over 600 members, both a Ladies section and a Juniors section and also a full fixture list of club competitions.

Right: For the last ten years one of the heralds of Spring has been the Reaseheath noticeboard advertising the Lambing Weekends. These hugely popular events attract thousands of visitors and for many families with small children they provide an introduction to the facts of life.

Agriculture students give up their time willingly to work during rhe Lambing Weekends. These two young men show off three new arrivals, including twins.

Right: Each summer since 1999 a maze has been formed at Reaseheath in a crop of forage maize, which is similar to sweetcorn. The plants grow up to three metres high, so visitors really do get lost in the maze. Everyone is given a flag to wave so they can be rescued if panic sets in. Each year has had a different theme and has proved hugely popular with children and adults alike. Pictured here is maze designer Adrian Fisher at Maze 2000, with its manager Richard Squire.

Royal visits and notable figures

Perhaps the most important aspect of a visit by Royalty or VIP is not the meeting of the illustrious person but rather the milestone the event creates in the memory.

Such visits become points of reference in the timescale of the college along with the opening of new buildings and the appointments and retirements of college Principals.

Reaseheath College has had only four Principals since it first opened in 1921, all of whom have made their unique contribution at the various stages of the college's development.

Over the years there have been so many true characters associated with the college that it would be impossible to mention them all. Two people from the early days of the college, however, made an indelible impression on all who met them: Willie Carr, the first Vice Principal, and Nellie Bennion of fearsome reputation, who made a formidable contribution to the teaching of cheesemaking in Cheshire.

W.B. Mercer

J.K Lamberton

G.J. England

V.J. Croxson

George V with Miss
J.Forster at the Worleston
Dairy Institute in 1913,
when it was honoured by a
visit from their Majesties
the King and Queen.

Above: In 1926 the Prince of Wales opened the women's hostel and the new dairy buildings. He appeared cold and irritable initially after being driven in an open car from Chester. No doubt his temper was restored by a cup of tea, and the college matron, Miss Jessie Wallis, kept his cup, unwashed, for the rest of her life.

Lord Leverhulme greeting girls sent to Reaseheath to train for the Women's Land Army in the Second World War. At the start of the war a poster campaign advertised for volunteers, but later women were conscripted into the WLA to help British farmers to produce the food that Britain so desperately needed.

Nellie Bennion advised on the design of the Cheshire School of Agriculture's dairy when the teaching of dairying transferred to the main campus from Worleston in 1926. She worked there for 27 years and almost every cheese farm in Cheshire and Shropshire had one of her students as its cheesemaker. Nellie Bennion was highly respected in dairying circles throughout the country.

Right: W.A.C.Carr was the first Vice-Principal of the Cheshire School of Agriculture. He became Vice-Principal in 1920 to Principal W.B.Mercer and the two men, both Scots, worked together to develop and stock the School, which opened in 1921. Mr. Carr also worked as a County Advisory Officer and with the War Agricultural Executive and later the Ministry of Agriculture Advisory Service. He retired to his native Scotland where he enjoyed a long and healthy old age, living to be a hundred.

Above: George England, Principal, in the mid 1970s with his Heads of Department: Ron Lawton, Fraser Nord, Ken Slater and Roy Kettleborough.

Princess Anne with George England chatting to dairy students at the official opening of the new Food and Dairy Technology buildings in 1982.

Opening the Lord Woolley Centre in 1987. Pictured from left to right are County Councillor Jim Brentnall (Chairman of the Governors), Vic Croxson (Principal), the Duke of Westminster and Lord Leverhulme.

Below: Princess Margaret opens the new milking parlour in 1991. She is pictured with Vic Croxson and Jim Brentnall. Seen in the background is W.A. Bromley Davenport, Lord Lieutenant.

Former Home Secretary Willie Whitelaw photographed at Reaseheath Awards Ceremony in 1995. Also pictured (left to right) are John Platt, Mrs. Whitelaw, Jim Humphreys and Vic Croxson.

The Board of Governors photographed in 2003. Back row from left to right: Rev. Peter Mascarenhas, Dr Nick Carey, Denis Parton, Tim Jones, Harold Bennison, Neville Care, George Hardy, Rev. Helen Chantry, Geoff Oakes(Clerk).Front row from left to right: Steve Wilkinson, Malcolm Rees, Penny Rudd, Vic Croxson, John Platt, Emily Thrane, John Dunning, Jane Clegg.

Wilfrid Bernard Mercer C.B.E, M.C., B.Sc., N.D.A..(1889 – 1962)

W.B.Mercer was born in Abbots Bromley Staffordshire, the third of four sons of the landscape painter Frederick Mercer. He was educated at Brewood Grammar School, Harper Adams Agricultural College and Edinburgh University. On graduation he was awarded a scholarship enabling him to do further postgraduate research at Rothampstead Experimental Station and Hamburg Botanical Institute.

In 1912 he was appointed Lecturer and then Senior Lecturer at Armstrong College, Newcastle on Tyne.

A career interruption came with his enlistment in the Warwickshire Yeomanry, with whom he served 1915 – 1918, seeing active service in Palestine and France. At Huj he was notably the only officer to survive an unsupported cavalry charge against Turkish field guns on a ridge 800 yards away, where finally swords carried the day.

1919 saw his appointment as Cheshire County Agricultural Advisor and Principal of the new Agricultural School being established at Reaseheath. He overcame the initial opposition of the local farming community with the aid of his distinguished and loyal staff.

During World War II Reaseheath became a centre for training girls for the Women's Land Army, and W.B. Mercer became Executive Officer for the Cheshire War Agriculture Committee, an exhausting and onerous duty, his being the final determining voice on dispossessing failing farmers. This experience undoubtedly shortened his life. He left Reaseheath in 1947.

His interests were wide. He was well known in Poultry Breeding, Grassland Research and Agricultural Education. He was appointed a University External Examiner and was Chief Assessor of a City and Guilds Committee standardising N.D.A.

examinations in all Agricultural Colleges, a post he filled until his death. He served as President or Chairman of various societies and in 1956 became President of Agriculture at the Bristol meeting of the British Association for the Advancement of Science.

W.B.Mercer was a modest academic who wrote and broadcast frequently, communicating with wit and authority. He was a keen sportsman, playing football and cricket with the students and appearing frequently with the Nantwich Cricket Team. He was also a Rotarian. He was a man who brought his own charm and grace into the Nantwich scene.

(Thanks to W.B.Mercer's daughter, Dr. Katherine Mercer-Young, for help with this information.)

James Kerr Lamberton B.Sc., N.D.A., N.D.D.. (1901 – 1976)

James Lamberton was born at Highburn, Fenwick near Kilmarnock on August 3rd 1901, the second son of John Lamberton, farmer. The family moved to Cocklebie Farm, Stewarton in 1903 and James attended the local school in Stewarton until leaving at the age of 14 to work full-time on the farm. In the late 1920s he decided to broaden his horizons, working on the dairy farm during the day and studying long into the small hours to gain his Higher Certificates. These qualifications gained him entrance to Glasgow University in 1930.

In 1934, following his graduation, he was appointed Assistant Lecturer in Dairy Husbandry at Leeds University.

During the Second World War he was Supervising Advisory Officer for the West Riding War Agricultural Executive and played a prominent role in developing wartime food production and in training members of the Women's Land Army. He was promoted to take charge of Copmanthorpe, an experimental farm attached to Leeds University, and also performed wardening duties there.

In 1947 James Lamberton was appointed Principal of the Cheshire School of Agriculture, Reaseheath, where he spent the next nineteen years until his retirement in 1966. During this time he he oversaw the expansion of training facilities at Reaseheath and the establishment of new departments in dairy and extra-mural sectors.

James Lamberton not only had a good working relationship with County Hall personnel, but was also respected by the local farming community in Cheshire. His years of experience working on the family farm were invaluable in this environment.

He was a member of the British Association for the Advancement of Science, a Rotarian and a sidesman at St. Mary's Church, Nantwich. He was co-founder and first President of the Nantwich and District Scottish Society. His passion, however, was violin making at which he was an acknowledged expert. One of his violins was sold recently at Sotheby's in London for a four-figure sum.

Always a modest man, he had a great sense of fun and enjoyed telling stories of his early years back home. He was a gentle man and a gentleman.

(Thanks to J.K.Lamberton's son, Andrew Lamberton, for help with this information.)

George J. England M.Sc., B.A. Cantab.. (1926 – 1995)

Born of farming parents George England was educated in Lancashire and later at Oswestry High School for Boys, where he captained both cricket and soccer first teams. He gained a scholarship to Selwyn College Cambridge and graduated in 1946. He later gained his M.Sc. from the University of Wales for research into animal husbandry and animal behaviour.

In 1948 he was appointed Lecturer of Agriculture at the Lancashire College of Agriculture, later becoming Farm Director and Senior Lecturer.

He was appointed Vice-Principal of Cheshire College of Agriculture in 1961 and became Principal on the retirement of J.K.Lamberton in 1966.

George England was Principal during a time of expansion and development and major new building projects at the college. During the 1970s extra student accomodation and lecture rooms were built. The existing Dairy buildings were turned over for different use and a new Food and Dairy Technology Department was equipped to a standard that made it the premier Food and Dairy unit in Europe at that time.

The Engineering and Horticulture Sectors were expanded and developed, as was the Extra-Mural department.

Student numbers grew during Mr. England's Principalship, as did their various activities. Sporting facilities improved from new courses in Turf Management and Golf Course Management and a nine-hole golf course was constructed. George England was Reaseheath Golf Club's first President and after his death a Sports Pavilion was built and named in his memory.

George England fostered excellent relationships with all aspects of agricultural activity in Cheshire. He was President of Cheshire Young Farmers for many years, a council member of Cheshire Agricultural Society and the first secretary of the Cheshire Grasslands Society. He also represented the college on many other committees.

George England's interests were in all things agricultural but he was also a keen sportsman and this continued into his retirement when he was instrumental in helping to establish a golf club for those who were retired either because of ill health or redundancy. He was a widely liked and popular man and remained active in many organizations up to his death.

(Thanks to George England's widow, Edith England, for help with this information.)

Vic Croxson D.L., M.B.A., N.D.A., F.R.Ag.S. (1946 –)

Vic Croxson was born in Falkenham, Suffolk, in 1946, the only son of a farmworker. He was an early product of comprehensive education, which he followed with two years of practical farming experience before taking a Diploma course at the Essex Institute of Agriculture (now Writtle) and a post-Diploma course at Harper Adams Agricultural College (now Harper Adams University College.)

Following a brief spell as a Research Assistant at Unilever's Pig Research Unit in Bedfordshire he was appointed as Lecturer and Warden at Shropshire Farm Institute (now Walford) in 1967. Responsibility for the hundred-sow unit on the college farm satisfied his early desire to be a pig farmer, but a passion for teaching had developed. He took an in-service course at the Wolverhampton Technical Teachers College (now Wolverhamplton University) and gained his Certificate in Education in 1974.

In 1978 a move to Staffordshire College of Agriculture (now Rodbaston) brought wider responsibility for curriculum management as well as for the famous Rodbaston Herd of Pedigree Large Whites – one of the oldest herds in the world. Promotion to Vice-Principal and Farm Manager followed in 1982 and following the sudden death of the newly appointed Principal in that year, an eight month period as Acting Principal.

Vic Croxson became Principal of Reaseheath in March 1987 and has led the college through the process of incorporation in 1993 and an unprecedented period of growth.

He has been a member of the Local Learning and Skills Council for Cheshire and Warrington as well as representing specialist colleges on the National Rates Advisory Group, the Learner Support Funds Working Group and a Specialist College Steering Group in which he fiercely

and successfully defended a ten percent specialist college uplift factor which was critical in enabling such colleges to remain financially viable.

Vic Croxson's services to Further Education have been recognized by the award of honorary membership of the City and Guilds of London Institute and freedom of the Guild of Educators. In 2003 he was appointed Deputy Lieutenant for the county of Cheshire.

A tireless committee worker and a man of great good humour and bonhomie, Vic Croxson takes early retirement from Reaseheath College in 2004.

Reaseheath on Show

There is never a day when Reaseheath is not on show. For staff, students and visitors the buildings and especially the grounds of the college bring pleasure and delight.

Each season presents a different beauty – thickly clustered daffodils and blossoms in the Spring, the green lawns and tranquil lake starred with water lilies in the summer, the burning red, russet and golden foliage of the Autumn and the tracery of bare trees against the cold pale skies of Winter.

At certain events during the year, however, the college puts all its energies into showcasing itself and its activities to the general public.

The Reaseheath Stand draws hordes of visitors annually at both the Cheshire County Show and the local Nantwich show.

The College Open Days have grown from occasions where local farmers could come to see examples of farming advances to family days out which attract thousands of visitors each year.

This early photograph from the 1930s shows farmers visiting Reaseheath for advice and ideas, on what was a forerunner of the modern Open Days. When the college became a business with the need to be financially viable these Open Days became the big family fun days which we see today.

During the summer months
the grounds of Reaseheath
College are open to the public,
and many visitors enjoy exploring
gardens such as this one.

Left: The first Principal of the college, W.B.Mercer, is photographed here with King George V, circa 1924, visiting the Cheshire School of Agriculture's stand at the Cheshire Show.

Nantwich Dairy Show in 1932, Nellie Bennion, Head of Dairying at the Cheshire School of Agriculture, is the only female judge.

Pictured at the Royal Show in Warwickshire in1995 is Reaseheath lecturer Tim Ball, about to take part in the national tractor driving competition.

Above right: Preparing for Tatton Park Garden Show in 2002 is Jonathan McDonald, Horticulture lecturer at Reaseheath College.

Below right: An interested audience at a Reaseheath Open Day looking for tips on successful container gardening from college experts.

Open Days provide an excellent chance for horses and riders to show off their skills in various competitions.

A Reaseheath student demonstrates the skill of dog grooming at the Cheshire Show 2002.

Left: June Shallcross, floristry lecturer, takes the Reaseheath College stand at the Nantwich Show.

Reaseheath stand at the
Cheshire Show in 2002.

A day of fun and festivities: visitors enjoy the sun on the lawns of Reaseheath Hall during Open Day 2003.

REASEHEATH: PICTURE OF A COLLEGE

The best of British – happy smiles from both girl and Bulldog after winning in a competition at Reaseheath Open Day 2002.

Is a little extra grooming
required? Cattle waiting to
be judged at a Reaseheath
Open Day.

Above: Reaseheath not on show: the first sad occasion when the college was closed to the public during the Foot and Mouth outbreak of 1967-1968. Staff watch as disinfecting precautions are taken at the college entrance. The college had to have its herd of cattle destroyed.

Left: Foot and Mouth disease strikes again in 2001. Vice-Principal Geoff Briggs oversees the precautions that happily prevented the college's animals from being affected by this outbreak.

Right: Open Day 2003 provides an outing for this Royal Python, one of the exotic residents of Reaseheath College's Animal Care Unit.

Left: Say it with flowers –
Reaseheath is the best.

Indian summer gives way
to Autumn, pictured here
at Reaseheath in 2003.

Always beautiful, whatever the weather, Reaseheath Hall is seen here in the snow.

Postscript

In 1922 James Worthington wrote the editorial of the Cheshire School of Agriculture's magazine "F.Y.M", of which the following is an extract:

"By the time this is in print we shall be on the eve of going down, most of us, alas, not to return. We shall all be very loath to leave Reaseheath. We personally are intensely grateful to our fate for leading our feet into such pleasant places and tender our heartiest thanks to a benevolent County Agricultural Committee. Of the need of such an Institution in our County there can be no possible doubt; even less of its practical use to such as ourselves.

.......Come yourselves, or send your sons, but in no carping spirit. If you do not learn anything, it will be because you know it all and are consequently to be envied. We personally have imbibed more knowledge of a useful marketable nature in twenty two weeks here than in the previous twenty two years."

These words express the sentiments of thousands of students over the years for whom the "Reaseheath Experience" has been one of the most precious times of their lives.